RWANDA

Threatened Kingdom, the Story of the Mountain Gorilla

This book has been produced on behalf of the International Gorilla Conservation Program to provide information about the challenges facing the survival of one of the most endangered animals on earth, the Mountain Gorilla. The book is intended to show how the social and historical context of the region affect the environment and how the environment has an impact on human livelihoods and the gorillas.

The format was chosen so that the information could be used across a broad spectrum of readers. The people who reside in the three countries where these unique animals live are resilient, hard working people who are committed to solving the problems they face. Over the past seven years they have gained my respect and admiration for a number of reasons, but primarily because of their determination to overcome adversity and to continue to provide hope for the future.

Howard G. Buffett

Threatened Kingdom

The story of the mountain gorilla

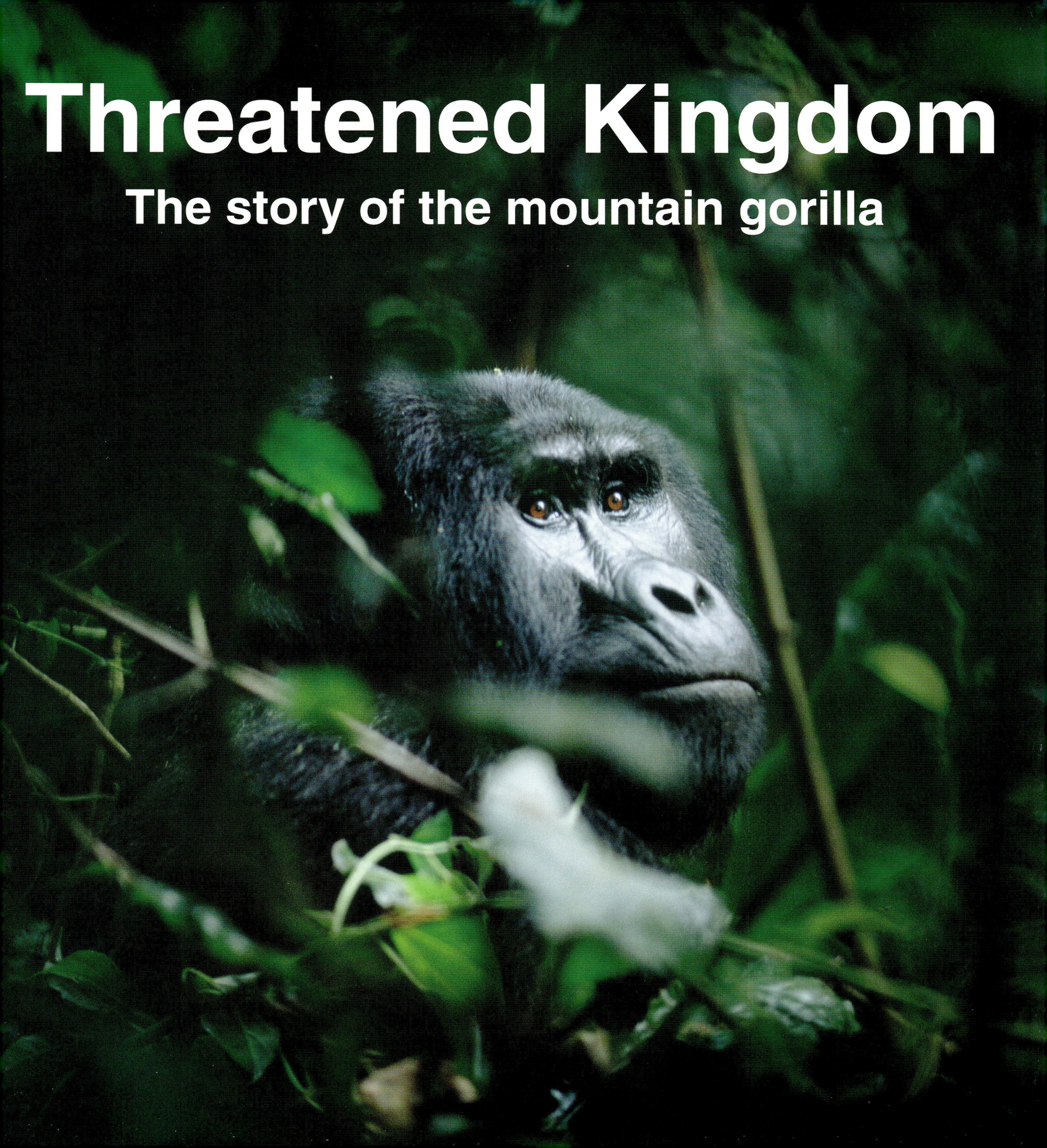

Dedicated to Mama Susie

Threatened Kingdom

The Story of the Mountain Gorilla

Published by

BIO Images

P.O. Box 4537
Decatur, Illinois 62525 USA

ISBN# 0-9707385-7-9

Library of Congress Card Number: 2004091042

Printed by Tien Wah Press (Pte) Limited
4 Pandan Crescent, Singapore

Photographs courtesy of Howard G. Buffett
Text and editing courtesy of Howard and Devon Buffett

Reprinted 2008

Production costs donated by The Howard G. Buffett Foundation

Foreword

by Annette Lanjouw

Mountain Gorillas are one of the rarest large mammal species in the world. Their small and fragile populations are found in a truly remarkable region of Africa....the western rim of the Great Rift Valley, where the drainage basin (watershed) of the Nile River is separated from that of the Congo River...the two largest rivers of the continent, and the second and third largest rivers in the world. This highland region is characterized by both dormant and active volcanoes, some of which are capped with snow, the aptly named "impenetrable forest" in Uganda and the misty Hagenia forests draped with dangling moss and twisted vines and orchids of the Virungas. These forests are home to the mountain gorilla families, led by the dominant alpha silverback males..... weighing up to 250 kg and protecting their harem families of females (much smaller than the males) and young. Young gorillas....little balls of black fur, with punk-rocker hairdos and perpetually astonished expressions...play tag, wrestling and tickling together while the adult males guard the family from unwanted intruders, and the females care for the infants, as vulnerable and dependent as a human baby.

This idyllic landscape, however, of peaceful gorilla families feeding and playing in lush montane forest vegetation, is so fragile. Surrounding these forests are enormous populations of people living in poverty and struggling to survive from day to day. The people, who have lived from the land for centuries have suffered enormously in the past 20 years from dictatorship, war and an almost total breakdown

of their social, economic and legal structures. Warring factions in a struggle driven by greed and social divisions have destroyed any chance that people may have a normal and peaceful life. For many, the only answer has been to turn back to the forest and to live off the resources that the forest and the land can give. Even before the outbreak of war in 1990, the steady growth in the human population has made it increasingly difficult for people to survive from agriculture and pastoralism, leading to degradation of the land and resources. This has led to competition over the limited natural resources remaining.....competition between people, and also competition between people and wildlife.

The gorillas, in their peaceful island of montane forest, are increasingly threatened through loss of their lush habitat from conversion of the land for agriculture, from deforestation for firewood and lumber, and from poaching and the transmission of diseases carried by humans to the primates.

Are we looking at the last years of their survival on this planet? Is this fascinating, intelligent and highly sociable, playful and gentle creature going to become extinct like so many other species in the world today? Not if current levels of conservation support can continue. Because remarkable as this may seem given the enormous challenges, this is one of the true conservation success stories. The population of mountain gorillas has slowly increased over the past 20 years. Their discovery by western scientists/explorers in 1902 initially led to a dramatic drop in their population...mainly from hunting for sport and poaching for the trade in live gorillas. Yet the efforts of African conservationists, together with support from Europe and the USA have led to very real and effective conservation. The reason this has been so successful is because the efforts have looked at the need for conservation in the context of other national and regional priorities. What people like to call "mainstreaming". Conservation cannot be just about the desire to protect a species from extinction. It must be about the need to preserve the environment, ecological functions and species, for both the sake of wildlife as well as people. If the conservation argument can be linked to economic arguments, cultural and social arguments, it has a much greater chance of success. The mountain gorilla, because it is exceptionally charismatic, but also tolerant of curious people who will pay for the opportunity to admire it, has become a tourist attraction in Rwanda, the Democratic Republic of Congo and Uganda, and is generating millions of dollars of revenue for those countries each year. In addition, it is starting to generate income for the poor rural communities who live around the gorilla's habitat. For these and other reasons, there are now real incentives to preserving these creatures, and their habitat. The people would be poorer without them.

It is important to remember that the world would be a lesser place without gorillas (and so many other species). You may not benefit directly from their conservation, but you do benefit from the diversity of wonderful creatures on this planet. To take care of ourselves, we must take care of the earth and its species. We are all connected, humans with every species of flora and fauna - each is part of the whole. A loss of one species is a tear in the fabric of human and natural life.

Preface

by Howard G. Buffett

At first glance, Rwanda, Uganda and Congo appear to share some of the problems that face many African nations. The people in each country are dependent primarily on subsistence agriculture and have gone through traumatic periods of conflict, which have left them struggling to survive. As in much of the world, the lives of many of the people are focused on their day-to-day survival and the long-term future is too distant to be of concern. This region is also unique, however, in that it is one of the five most important regions of the world in terms of biodiversity and species endemism. And it has a special resource that is found nowhere else in the world: the Mountain Gorilla.

The Virunga volcanoes that separate the Democratic Republic of Congo, Uganda and Rwanda, and Bwindi Impenetrable Forest were first discovered by British explorers in 1861. It was later, in 1902, that the gorillas were discovered. However, the Mountain Gorilla species was made famous through the research of Dian Fossey, and today this species' survival depends more on the local people of these countries than any other factor. The question remains whether local human needs can be met while saving the critical habitat these animals need for survival. It is the only place on earth where Mountain Gorilla occur naturally.

Pressure for land and years of conflict have put these gentle giants in the crossfire. The need for income has put the gorilla at risk from illegal trafficking in young gorillas, and from poaching for food. Although gorillas are not generally eaten in this region, snares set for other animals can inflict serious damage. A gorilla can lose a hand or a foot to a snare, with the subsequent risk of infection and death. The constant pressure on the gorillas has taken a high toll over the years.

Fortunately, through the incredible commitment of dedicated park rangers, increased understanding by local inhabitants and outside resources from interested organizations, the Mountain Gorilla population has become reasonably stable over the past few years. However, even the loss of a few animals can have a significant impact on the population. With such low numbers (currently estimated at approximately 700 gorillas) disease, habitat encroachment, and even limited indiscriminate killing of these animals can have devastating results.

Hope lies in understanding the local challenges and providing solutions to the needs of the people who must live alongside these highly endangered primates. Often outsiders make judgments about what has more value; gorilla habitat, local needs, or commercial development, yet outsiders do not struggle with the day-to-day challenges faced by the people who must share the same resources. Therefore, it is the generations of Rwandans, Ugandans and Congolese who must make daily sacrifices that will ultimately decide the fate of this entire species. I hope this book provides insight into the circumstances and challenges that impact this species. If you ever have the opportunity to look into the eyes of a gorilla, you will instinctively know that these unique creatures deserve our efforts to save them.

Bwindi Impenetrable Forest, Uganda.

The Volcanoes National Park, Rwanda (Right). The two forest blocks (Bwindi Impenetrable Forest and Virunga Forest, comprised of the Volcanoes National Park in Rwanda, Virunga National Park in the D.R. Congo and Mgahinga Gorilla National Park in Uganda) are home to the last remaining Mountain Gorillas in the world. The Bwindi Impenetrable National Park covers 321 square kilometers (124 square miles) and the Virunga Forest 444 square kilometers (276 square miles) of forest. The Virunga forest includes a chain of six dormant volcanoes. (Nearby, the rest of the Virunga National Park, covering 7,800 square kilometers 4847 square miles), contains two active volcanoes.

Goma, the largest urban center in DRC, is located near the mountain gorilla habitat and is built on the foothills of the active Nyiragongo volcano (Right). The eruption in 2002 wiped out more than 50 percent of the town, leaving hundreds of thousands of people homeless. Rebuilding those homes has led to extensive deforestation, with inevitable impact on the forested parks.

Available resources are used for local needs, stripping the forest directly up to the park borders. The rich volcanic soil of the Virunga Range is considered a valuable asset for crop production. Therefore, there is usually no buffer zone remaining between the animals in the park and human activity (Left: Bwindi Impenetrable Forest; Uganda; Right: Foothills bordering the Volcanoes National Park; Rwanda).

The technique used to clear virgin forest for agriculture is commonly referred to as "slash and burn" agriculture. Over 95 percent of the crops grown in this region are subsistence crops, grown for local consumption rather than for sale or commercial use.

Like a patchwork quilt, little squares of farmland cover the steep slopes of these hills providing a living to thousands of local farmers (Rwanda).

Valley after valley, the original home range of the Mountain Gorilla has been converted for human use. Terraces provide a way to utilize almost all available space for crop production, while trying to control soil erosion (Uganda).

Commercial production also encroaches on wildlife habitat. (Uganda) The development of tea and coffee plantations has contributed significantly to increased foreign revenue. Some believe these commercial activities should take precedence over saving animals and habitat (Right).

Clearing usually takes place with machetes or axes; little heavy equipment is available for land clearing. With complete tree removal difficult, a tea field is planted around the remaining stumps (Below).

Agriculture plays a key role in rural areas. Crops such as bananas, beans, carrots and potatoes provide food and a small amount of income for local people. The majority of these crops are used for immediate consumption, the balance is transported to urban areas and sold in local markets.

Mountain Gorillas rely completely on the forests, which in the past were converted or consumed for human use. Gorillas roam throughout their forest habitat and sometimes cover large distances in a day. Although gorillas are not strictly territorial, they do have home ranges which can be as large as 40 square kilometers (15 square miles). A primary reason for such an extensive range is the need for large amounts of vegetation. Gorillas move regularly in a constant search for food, thus allowing vegetation to regenerate in their absence. Large uninterrupted areas of habitat are therefore required to support gorillas.

Gorillas build their nests for sleeping at night. They use leaves and other plant material to make these one-night resting spots. Researchers can estimate the size and age structure of a group by how many nests are constructed and the size of the feces by the nest. As infants share a nest with their mothers, these nests will have large and small feces next to them. However, such indirect counting methods are not as accurate as visual observations of the group.

A healthy forest provides benefits to both gorillas and humans. This includes maintaining the conditions for a viable watershed, which in turn provides clean water. The tree canopy of the forest slows the impact of heavy rains, and combined with the roots of trees and plants, prevents erosion and sediment from polluting fresh water supplies. Proper forest management preserves the resources needed for both humans and animals (Munyaga River, Bwindi Impenetrable Forest).

A few kilometers downstream outside the park border, the Munyaga River is used for washing clothes, bathing and boiling water. Over-cutting of the forest can threaten future availability of adequate clean water supplies (Right).

Young women approach the park border in Rwanda to collect water for cooking. Water supplies closest to the forest tend to be the cleanest, since the forest acts like a giant filter for runoff water (Below).

A wall has been constructed along the border of the Virunga Volcanoes and Mgahinga Gorilla National Parks, to prevent large mammals such as elephant and buffalo from coming out of the park and damaging people's crops. When animals wander out of the park, they can totally destroy a local farmer's crops in just a few hours.

Cape Buffalo (Below).

Productive soils and adequate rainfall make the foothills adjacent to the parks ideal for the production of bananas and other crops which are an important source of cash income.

The forest provides a number of resources for local communities. Trees from both inside and outside the parks are used for lumber (above) and firewood (next page); vegetation is used for animal feed and shelters (page 43). Maintaining a healthy forest is important to sustaining a productive watershed, which in turn provides important resources, such as medicinal plants, better soil fertility and buffers for potential diseases to domestic animals and humans.

A cow stands among a crop of banana trees and children play in a banana grove. With the increase in agriculture surrounding the gorilla habitat, the quiet and peace of the Mountain Gorillas' forest, is sometimes shattered by the noise of a cow or the voices of children.

Much of the produce raised in the area is sold at local markets to provide cash for every day living needs.

For produce not used in local markets, farmers prepare their crop to be transported to larger urban areas. There are few economic alternatives to replace farming in this part of the world.

Mountain Gorillas are protected in four national parks in three countries. The national park authorities of Rwanda, Uganda and Democratic Republic of Congo, coordinate their activities and monitor wildlife and illegal human activities throughout the gorillas' habitat. (Above, park employee of Democratic Republic of Congo).

Loaded down with charcoal and a child, this woman uses any available resources she can acquire. Although electricity is available in many urban centers, it is not distributed into rural areas, and many urban people are too poor to own electrical appliances. As a consequence, more than ninety-four percent of the population in this region is entirely dependent on fuel wood and charcoal for all their cooking and heating. This puts enormous pressures on the forests and leads to extensive deforestation within the national parks.

A young boy hauls bamboo and tree limbs from the forest.

A silverback observes his kingdom of bamboo, a key resource which humans extract from the forest (Opposite page).

This region of Africa is exceptionally diverse in wildlife, ecosystems and habitats. Other wildlife found in the region besides the mountain gorilla include a range of primates, such as the olive baboon (Below) and the vervet monkey (Right).

Chimpanzee (Right).

Black & White Colobus Monkey (Below).

Elephant (Right) range both in the forest and the drier savannas of this region, as do many ungulates, such as Waterbuck (Below).

Gorillas often exhibit various poses that appear comical and sometimes they seem to mimic human behavior. They share many characteristics with humans, including a high degree of intelligence, complex social behaviour, strong emotional bonds and well-developed senses (sight, smell, taste and touch).

An adult gorilla will consume up to 20 kgs (44 pounds) of vegetation in a day, although a silverback may consume as much as 35 kg (77 pounds) per day. Therefore, enough space and a healthy forest are critical for the gorillas' survival. Encroachment on the gorillas' habitat will reduce the available food sources and therefore increase the chance that this species could be lost forever.

Gorillas are the largest primates. An adult female gorilla will usually weigh between 70 and 120 kgs (154 and 265 pounds). An adult male will average about 160 kgs (353 pounds), although a male can reach as high as 227 kgs (500 pounds).

The region encompassing the Mountain Gorilla habitat has experienced years of conflict. The home of the Mountain Gorilla has been the location of political struggles involving a large number of countries, including the Democratic Republic of Congo, Rwanda and Uganda. The conflict has had a devastating toll on the human population, with over 900,000 people killed in the brutal genocide in Rwanda and up to 3.5 million people killed in the DR Congo. Millions of people have been forced to flee their homes and live in temporary shelters, living off the land. As families were split apart and children orphaned, the rest of the world did little to stop the violence. The war also impacted the environment, and the wildlife that is unique to this region. Hundreds of lowland gorillas were killed, and it is estimated since 1994, between 18 and 22 Mountain Gorillas were killed as a result of the fighting which took place. Although this number is low compared to what it could have been, it represents almost four percent of this fragile population.

The war and genocide in Rwanda involved the systematic killing of almost one-million people. The impact of this period in Rwanda's history will last for generations and has reverberated across the continent. In order to face, learn from and avoid a repeat of such horror, memorials have been erected at several genocide sites.

At Murambi, (Right) bodies are preserved with lime, the purpose is to remind the world of what happened; and to never let it occur again.

A church in Ntarama (Below) shows the remains of a few of the hundreds of massacred bodies from the 1994 genocide in Rwanda. Dancilla Nyirabazungu sits in the corner of the church and recounts her story of how she barely escaped with her life. As tragic as these deaths, was the fact that no intervention occurred as the world watched these events unfold.

The roof of this shelter is an old United Nations tarp, evidence of a decade of displacement and destruction (Right).

It is estimated that 600,000 refugees returned to Rwanda between 1995 and 1996. Camps for returned refugees and internally displaced people were established in the Gishwati Forest Reserve, resulting in the obliteration of this forest for firewood and agricultural land. The forest was completely depleted in only four years. Today, many of the "temporary" settlements have become permanent. The forest that once provided wood for shelter and cooking is now exposed and experiences erosion and dramatic mud slides in the rainy season (Inset).

The United Nations High Commissioner for Refugees provided much needed support to the refugees and other displaced people during the conflict in the Great Lakes Region. Populations that lost their homes, their belongings and fled fighting or volcanic eruptions received food and shelter from the UN. The large concentrations of homeless people placed additional pressure on natural resources, leading to increased loss of forest cover and wildlife.

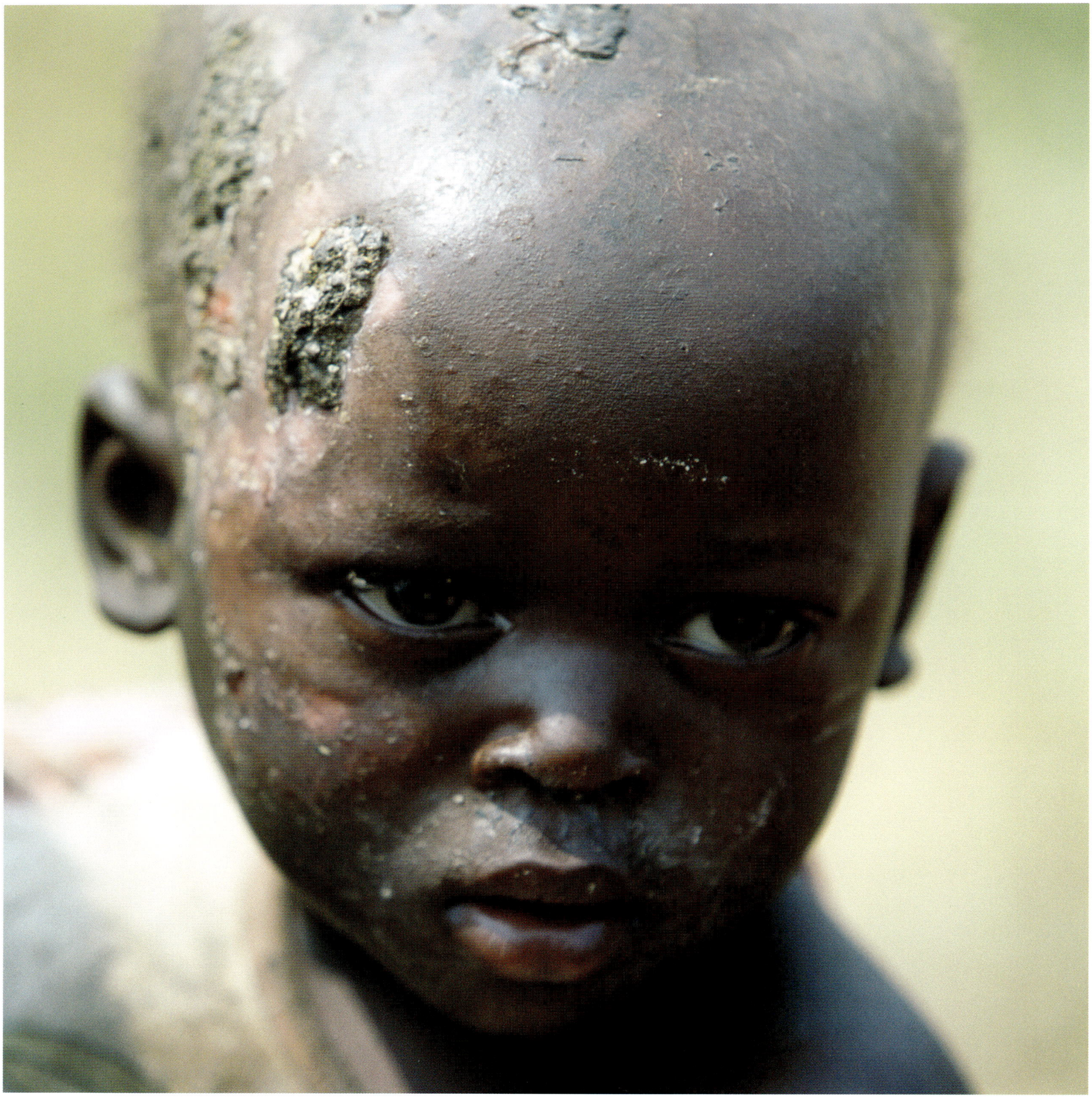

Years of turmoil have stretched resources past their limits. Many people in the rural areas cannot afford and do not have access to adequate medical care.

Over 300,000 children lost one or both parents as a result of the war and genocide and ensuing displacement of local populations.

The International Gorilla Conservation Program (IGCP), with assistance from outside resources, has developed a comprehensive plan to help provide economic support for local communities while protecting the Mountain Gorillas and their habitat.

Involving the local populations is part of the conservation strategy for the Mountain Gorillas. This involves jobs such as guides, trackers and porters for gorilla tourism, ownership and management of tourism facilities and helping to monitor the gorillas and their habitat. A portion of the fee for gorilla tourism is shared directly with the local communities to fund development projects in the region In the past, prior to the war, gorilla tourism was the third-highest generator of foreign exchange in Rwanda. It is recognized by all three counties that revenue from gorilla-viewing can contribute significantly to the economy of the region.

The development of economic alternatives for local people helps to reduce pressure on the forest. Beekeeping is a compatible profession that meets this objective.

Local talent comes in many forms. Here a tourist views drawings created by a young Rwandan.

Wood carvings, baskets and other locally made products are sold to tourists. This income can help replace profits derived from the natural resources found in the gorilla's habitat (Right).

Local culture is shared with tourists in the evenings around a campfire. Employment through lodging and support services is an additional way that local people can benefit from gorilla tourism (Below).

THE COUGAR FUND
Protecting America's Greatest Cat

Two juveniles wrestle (this page) while an adult stretches (opposite page). It has been documented that gorillas use at least 25 recognized vocalizations including grunts, roars and growls.

A baby mountain gorilla is dependent upon its mother for about three years. At about two months a baby gorilla can crawl and at about four months of age, it will begin to ride on its mother's back, clinging to her fur.

This baby Mountain Gorilla will nurse for about eighteen months to two and a half years. A new-born gorilla weighs about 1.4 to 2 kgs (3 to 4 pounds); this is about half the weight of a human at birth. Young gorillas are very vulnerable and will stay close to their mothers for protection.

Gorillas, like all primates, prepare their food before eating it. The tough and sometimes bitter bark or leaves are often stripped off the stems of plants, so that the tender core can be enjoyed. Gorillas use their fingers and their teeth to delicately remove the parts of the plant they do not like to eat, and only indulge in those parts that provide them with the most nutrients or water.

Mountain Gorillas eat up to 70 kinds of plants and as many as 200 different parts of various vegetation. This includes wild celery, leaves, herbs, thistles, nettles, shrubs, roots, bark, flowers, fruit, fungi and bamboo. Sometimes gorillas will eat ants and termites to supplement their diet.

Gorillas eat the leaves of thistles, but also the thick base and the roots. Roots, like leaves and stems, add a variety of minerals and nutrients to a gorillas' diet.

Dead trees decay slowly in the forest, collecting humidity and many minerals. Gorillas will sometimes chew on and eat the decaying wood, as well as fungi that grow on the decaying trees. The soft wood is a source of nutrients and water, particularly valuable in the dry season, when water becomes scarce. The order in which gorillas are allowed to enjoy this wood is directly related to their order in the hierarchy in the group, with dominant animals having first access.

Males mature around the age of ten to twelve years, and about that time the hair on their backs begins to turn silver. Therefore, the mature males are referred to as silverbacks. The silverbacks are the leaders of the family group.

There are two species of gorilla, and four sub-species. The Mountain Gorilla, Gorilla beringei beringei is the most endangered, together with the Cross River Gorilla, Gorilla gorilla diehli. The Mountain Gorilla population numbers approximately 700 individuals. The Western lowland gorilla, Gorilla gorilla gorilla, and the Eastern Lowland Gorilla, Gorilla beringei graueri are more abundant and can also be found in zoos (pages 98-101). No Mountain Gorillas are found in zoos, primarily because they have never survived in captivity.

One of the most obvious differences between the three gorilla sub species is the Mountain Gorilla's longer hair. This is primarily because it lives at higher altitudes, ranging from 1,600 to 4,000 meters (5,249 to 13,123 feet) and needs the hair for warmth. Mountain Gorillas also have larger jaws, smaller noses and shorter arms than their relatives.

The only gorillas that exist in zoos are western and eastern lowland gorillas. Attempts were made to collect mountain gorillas for zoos in the 1960's and 70's, but none survived. Since then, all zoos have agreed to halt all further attempts to collect mountain gorillas for captive populations, to avoid making the wild population even more vulnerable to extinction.

Captive lowland gorillas in zoos are important ambassadors for the conservation of wild populations of gorillas, and both lowland and mountain gorillas benefit from the impact a gorilla has on the people that see them in zoos. Although mountain gorillas are slightly larger and have longer fur than lowland gorillas, the differences are relatively slight and most people would have difficulty identifying the difference between them.

Generally, the gorillas are peaceful and non-aggressive animals. However, the silverback is famous for his ferocious nature when he is threatened or when protecting his family (Right).

In spite of his potential aggression, the silver-back can be very tolerant and even playful with youngsters (Below).

Mountain Gorillas have been divided into three categories: "wild" gorillas, "research" gorillas and "tourist" gorillas. Wild gorillas essentially have no human contact; research gorillas are approached and observed strictly for research purposes; and tourist gorillas are habituated to tolerate short visits by people to generate income for conservation and community development.

Gorillas' wounds can be inflicted by humans or by fights with other gorillas. Open wounds pose threats of infection and even small cuts can turn into serious problems.

Young gorillas are playful and like to climb. As they get older, they spend less time climbing and more time eating. Gorillas spend about 30 percent of their time eating, 40 percent sleeping and 30 percent traveling.

Adult gorillas have 32 teeth. They use their large molar teeth to chew food and long canine teeth to bite or break thick pieces of vegetation.

RWANDA

Gorillas' hands and feet are very similar to ours. They have five fingers and five toes, including a "big" toe. Gorillas can grasp objects with both their hands and their feet.

Each gorilla's nose is as unique as a human fingerprint. This is one way that researchers identify individual gorillas (see pages 114 and 115).

Sample Research Card

Sample Research Card

Information about the International Gorilla Conservation Program

International Gorilla Conservation Programme

The International Gorilla Conservation Program (IGCP) operates in a Landscape context, focusing on key conservation targets and working with a variety of different people and groups to reduce the threats to conservation. These people include the park authorities, local and regional governments, local people and environmental experts, all of whose activities have an effect on the environment and the natural resources in that landscape.

The strategy of IGCP is based on three axes:

- *Establishing a strong information base to allow decision-makers to understand the dynamics between the human population and the natural habitat/wildlife.*
- *Strengthening the protection of the habitat and mountain gorillas through regional collaboration by the three countries and structured mechanisms for transboundary conservation;*
- *Reducing threats to the conservation targets by assisting the human population in developing livelihood strategies that are complementary to, and even contribute to conservation objectives.*

Effective conservation of endangered species and habitats has become increasingly complex in this part of Africa due to the continued political instability of the region. IGCP has adopted an approach which allows it to respond in a flexible manner to the needs and priorities on the ground.

The goal of IGCP is the long-term conservation of the mountain gorilla and its regional high-and medium altitude forest habitat in Rwanda, the Democratic Republic of Congo and Uganda.

The International Gorilla Conservation Program (IGCP) was formed in 1991, bringing together three international conservation organizations and three protected area authorities: the African Wildlife Foundation (AWF), Fauna and Flora International (FFI) and the World Wide Fund for Nature (WWF); the Institut Congolais pour la Conservation de la Nature (ICCN) in the Democratic Republic of Congo, the Office Rwandais de Tourisme et des Parcs Nationaux (ORTPN) in Rwanda and the Uganda Wildlife Authority (UWA) in Uganda.

The program is based on the experience and long-term involvement of these organizations in conservation work in Africa and elsewhere. IGCP has focused on bringing together the joint expertise of African conservation professionals with experts from many different parts of the world, both as part of the team and as partners on the ground.

One of the principal characteristics of the program is that it is regional, basing its activities in the afromontane forest habitat that lies across the shared borders of Rwanda, Uganda and the Democratic Republic of Congo (DRC). Emphasis is given, at all levels, on regional collaboration, to ensure that the three countries work together to protect and conserve this area.

Regional activities include: ecological monitoring and surveillance; ecotourism and nature-based tourism development; joint training, communication and sharing of experiences; planning; community participation and conservation enterprise development, as well as park management planning.

For further information about the program, please use the following contact information:
www.mountaingorillas.org
www.igcp.org

IGCP
Britak Centre
Mara Road
P.O.Box 48177
Nairobi, Kenya
Tel: 254-20-2710367
Fax: 254-20-2710372
igcp@awfke.org

Map courtesy of Scott Zillmer

Organizational Contacts

The African Wildlife Foundation (AWF) recognizes that the wildlife and wild lands of Africa have no equal. We work with people - our supporters world wide and our partners in Africa to craft and deliver creative solutions for the long-term well being of Africa's remarkable species, their habitats and the people who depend on them.
For more information, visit www.awf.org

AFRICAN WILDLIFE FOUNDATION

CONSERVING WILDLIFE. RESPECTING ALL LIFE.

Fauna and Flora International (FFI), founded in 1903, is the world's longest established international conservation body and one of only a few who's remit is to protect the entire spectrum of endangered species of animals and plants worldwide.
FFI's mission is to conserve threatened species and ecosystems worldwide, choosing solutions that are sustainable, based on sound science and take account of human needs.

For more information, visit www.fauna-flora.org

FAUNA & FLORA
International
Conserving wildlife since 1903

WWF is one of the world's largest and most experienced independent conservation organizations. Its mission is to stop the degradation of the planet's natural environment and to build a future in which humans live in harmony with nature, by:

- *Conserving the world's biological diversity*
- *Ensuring that the use of renewable natural resources is sustainable*
- *Promoting the reduction of pollution and wasteful consumption*

For more information, visit www.panda.org

WWF

Acknowledgements

We would like to acknowledge the dedicated and courageous efforts of the authorities in the three protected areas where the world's only mountain gorillas exist. It is a result of their work that these highly endangered primates still survive today. During the past 10 years of war, many of the park staff have lost their lives or those of family members, and they deserve the recognition and gratitude of the conservation community. Despite the difficulties and challenges, the park rangers and wardens continued their work without interruption. The few hundred staff of the Rwanda ORTPN, Congolese ICCN and Ugandan UWA have conducted countless hours of surveillance and monitoring patrols and worked with local communities to help them improve their lives. Their efforts have helped to achieve maximum conservation impact. In spite of all the challenges they have accomplished what could not have been expected: a halt in the encroachment of the forest and an increase in the population of mountain gorillas over the past 10 years.

We would also like to acknowledge the many supporters who have contributed generously to conservation of the mountain gorillas, thus enabling this valuable work to continue.

A special thanks to the following individuals who assisted in the preparation and publication of this book: Mbungo Antoine, Médiatrice Bana, François Bigirimana, Melinda Binks, Déo Kajuga Binyeri, François Bizimungu, Derek Books, Howard W. Buffett, Cynthia Clemens, Tony Clouston, Jim Doherty, Maryke Gray, Scott Hawbaker, Peter Kinnear, Annette Lanjouw, Bob Little and Allied Photocolor, Pétillion Lutuya, Matthew Mangelsen, Tom Mangelsen, Nsabimana Marc, Kayembe Masumbuko, Déo Mbula, Antoine Mbungo, Emmanuel Mugenzira, Benjamin Musisi, Mark Mwine, Joseph Ngango, Justin Nyampeta, Dancilla Nyirabazungu, Thiele Robinson, Rosette Rugamba, Pierre Ruhumuliza, Paulin Wilondja Ngobobo, Scott Zillmer.

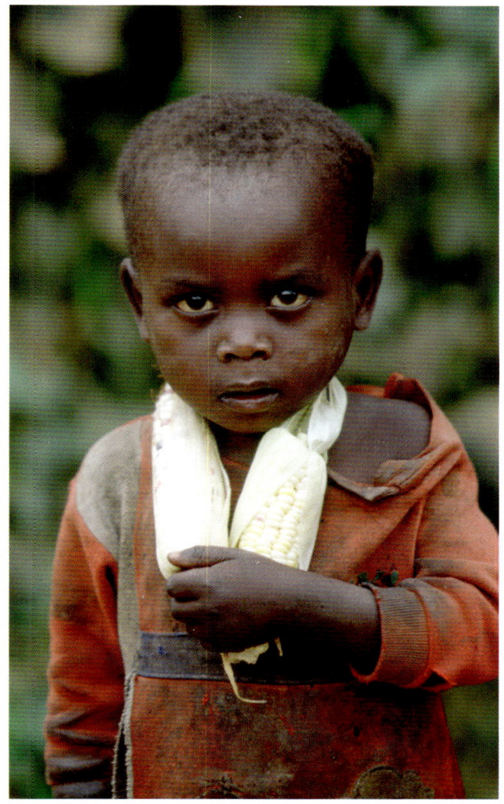

Will man prevail over nature? The way we use our power will determine whether or not our conscience truly makes us a better steward of all species.

As we devour precious resources on a worldwide basis, will our actions force species such as the Mountain Gorilla into smaller and smaller areas? Will we slowly destroy the only remaining habitat these unique creatures depend on for survival?

As if stranded on an island, these animals have nowhere to run, nowhere to hide. Their only true predator is man.

If history is any indication, time will eventually run out for the Ngagi, the local name for Mountain Gorillas. However, the commitment of the people who have given their lives to save these animals provide hope that the story will not end. My dream is that others can experience the magic of sitting, watching and listening to these incredible animals for many years to come.